Do Wolf Pups Need a Babysitter?

by Mary Clare Goller

Do wolf pups need a babysitter?
Yes!

What else do wolf pups need?

Wolf pups need shelter.

A hollow log is good shelter.

5

Wolf pups need air.

They breathe in air.

Wolf pups need water.

They drink water from a stream.

Wolf pups need food.

They get nutrients from food.

Wolf pups have basic needs.
What are they?